DETERGENT
PACKAGING INNOVATIONS

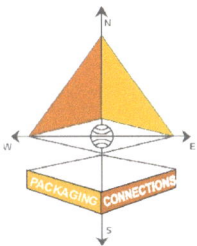

Sanex Packaging Connections Pvt. Ltd.
www.packagingconnections.com

Copyright

Published by :

Sanex Packaging Connections Pvt. Ltd.

An ISO 9001 : 2008 Certified Organisation

117, Suncity Tower, Sector-54

Golf Course Road, Gurgoan-122 002.

Tel : +91 124 4965770

Fax : + 91 124 41433951

e-mail : info@packagingconnections.com

Like us on Facebook : www.facebook.com/pconnection

ISBN :978-9385010026

List of Contributors

Team www.PackagingConnections.com by Sanex Packaging Connections Pvt Ltd

Sandeep Kumar Goyal, Founder & CEO
Amita Venkatesh Valleesha, Associate: Scientific Affairs & Consultancy
Chhavi Goel, Associate: Research & Business Consulting
Bhaskar Ch, Technology Advisor e-business
Rohit Raj Sachan: Executive Technical
Kuldeep Karan, Graphics/Web Designer

DETERGENT PACKAGING INNOVATIONS

Table of Contents

DETERGENT PACKAGING INNOVATIONS

Introduction

Idea behind this book is to bring the innovations to wider group of professionals to meet the mission of packaging knowledge sharing and that too cost effectively. We feel that this publication will further fill the project pipelines of companies and improve the standards of packaging. Many professionals either do not have the access or time to go through so many innovations together. So we think this publication will fill that gap. For your feedback please email directly to info@packagingconnections.com

With this, Enjoy Wonders Of Packaging!

Sandeep Kumar Goyal
Founder & CEO ,
www.PackagingConnections.com

DETERGENT

It is a water soluble cleansing agent that combines with impurities and dirt to make them more soluble, and differs from soap in not forming a scum with the salts in hard water. This may be in form of solidified bar, loose powder, thick liquid or concentrated balls

GLOBAL and INDIAN Detergent Market Scenario

Globally total detergent market size is estimated at 13,745 million Euros that covers 48% market share in all detergent and maintenance products category (including home care products like floor cleaners).

In this powder detergent contributes to 3,411 Mn Euros that covers 24.8% market share. Liquid detergents of 4,238 Mn Euros with 30.83% market share. Unit Dosage is 936 Mn Euros with 6.8% market share. Fabric conditioner is 2,335 Mn Euros with 16.98% market share.

(Source- Euro Monitor International)

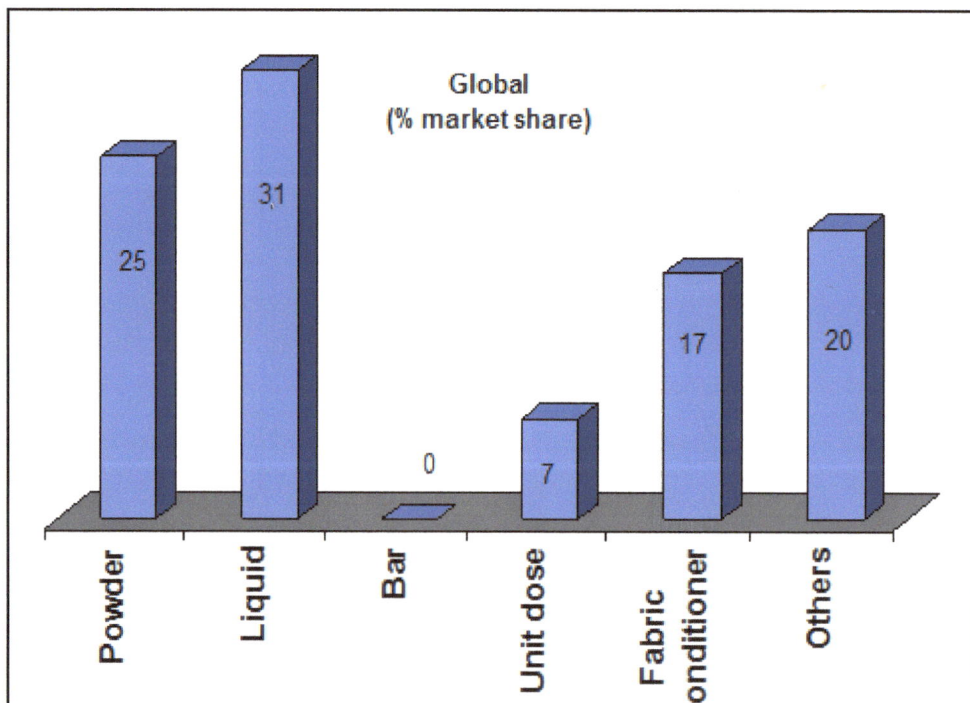

DETERGENT PACKAGING INNOVATIONS

With increasing modernization, demands of more comfort products are increasing. Laundry care industry has been growing @ approximately 11% in India and is forecasted to reach 241 billion by 2017. Globally it is growing @ 9%.

In terms of value - Indian detergent market is estimated to be 5100 Cr (1.2 bn USD). Detergent Bar market share is 43%, while liquid detergent is 12%. Powder detergent is 42% which has now increased in metro cities.

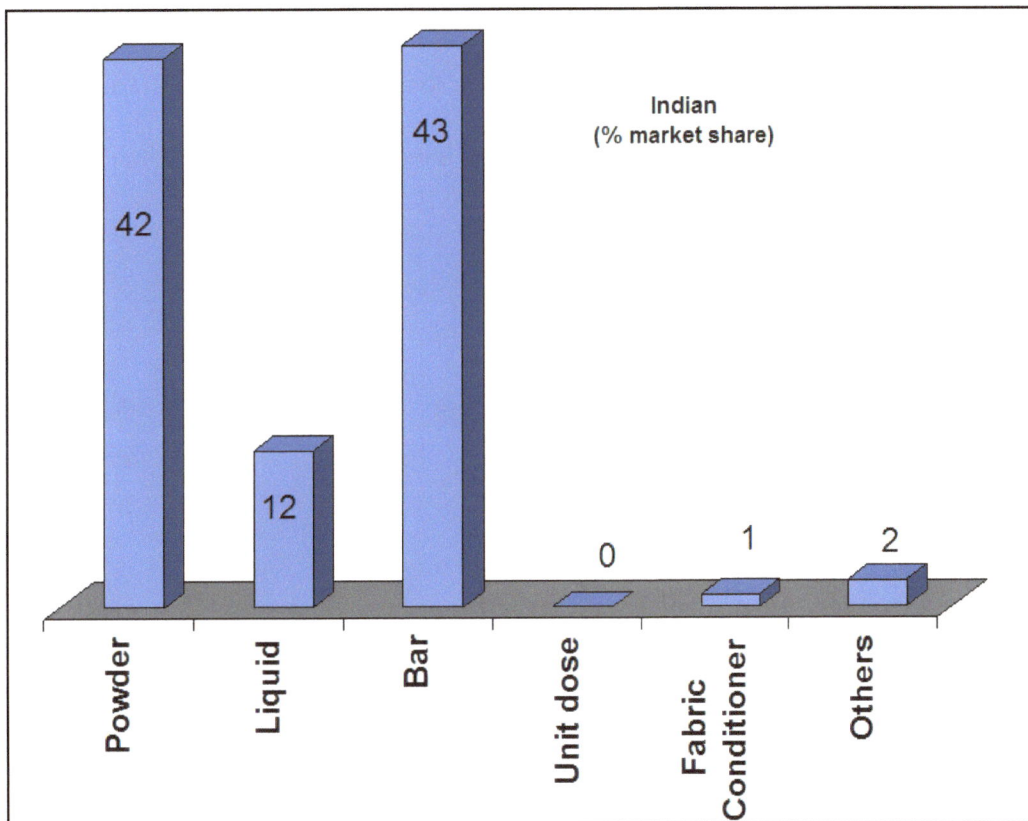

With the increasing sales of washing machines, liquid and powder detergent share is also increasing. Powder deter¬gent is the most common form being used since this can be used for hand wash as well as machine wash purposes. With the modernization, liquid detergent is anticipated to grow especially in Urban areas.

Unit doses is one category of detergents that has shown high growth in developed countries but there are yet to flour¬ish in developing nations. Multinationals have come up with this concept to give eco-friendly packaging options to detergents thereby reducing waste.

DETERGENT PACKAGING INNOVATIONS

Over the last few decades, India has grown tremendously and so have the incomes of its people, with a larger chunk of the Indian demography moving into urban areas in search of better and more lucrative career options. With this advent, there has been a significant rise in consumer durables as well, with some of the earliest signs of this rise being shown by those which fulfilled the most basic of needs, like washing clothes. As a result of this, there has been an in¬flux of buyers in the detergent market, with more and more people buying washing machine powders. However, this constitutes only some % of the total population.

Indian detergent market is divided in to 2 parts according to their washing style-

a. Hand Washing (82%) — Prevalent ones are Powder Detergent & Detergent Bars

b. Machine Washing (18%) — This involves mainly Powder detergent & Liquid Detergent

However globally scenario is different where there is no or rare concept of hand washing; only machine wash is available so mainly powder and liquid detergents are prevalent.

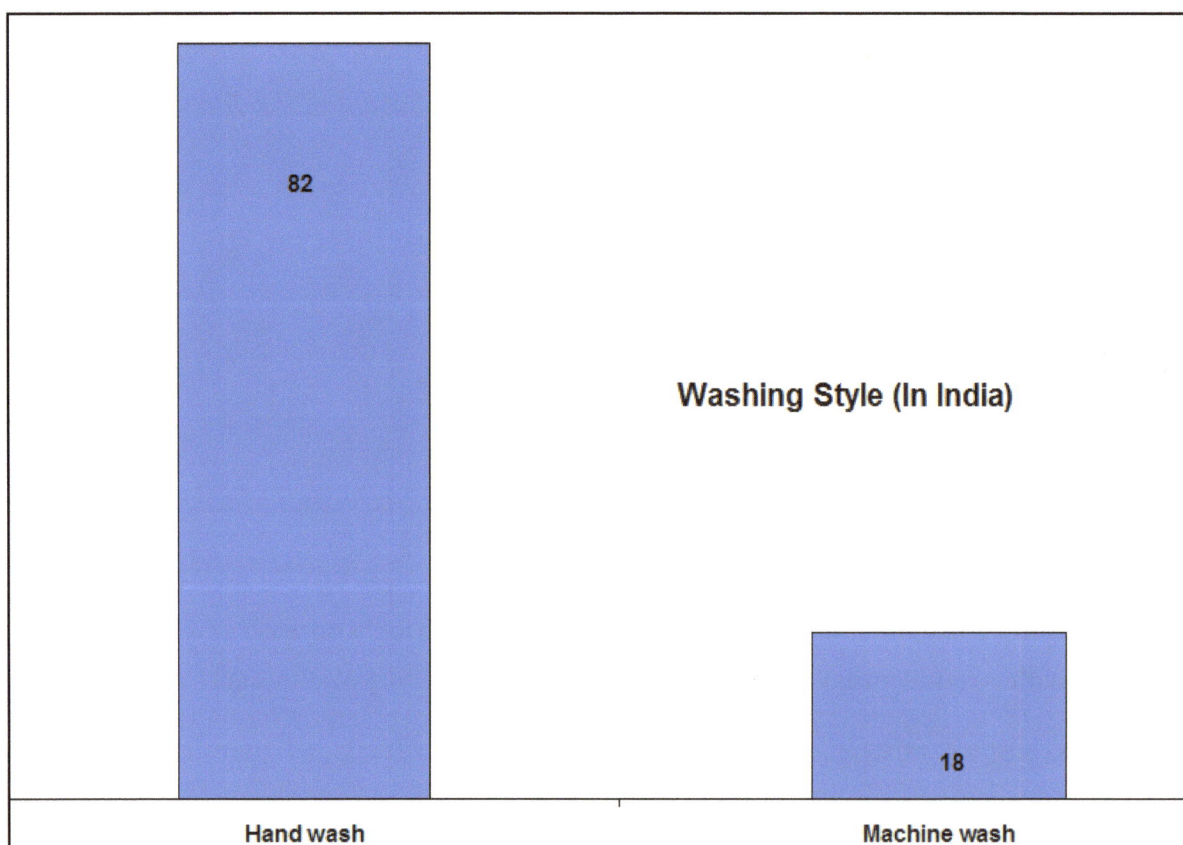

DETERGENT PACKAGING INNOVATIONS

Per capita consumption of detergent in India has been estimated 2.7 kg annual (Rural areas) and in Urban areas is 3.9 kg annual. Globally it is 4.9 kg (Asian Countries) & 10 kg (European Countries).

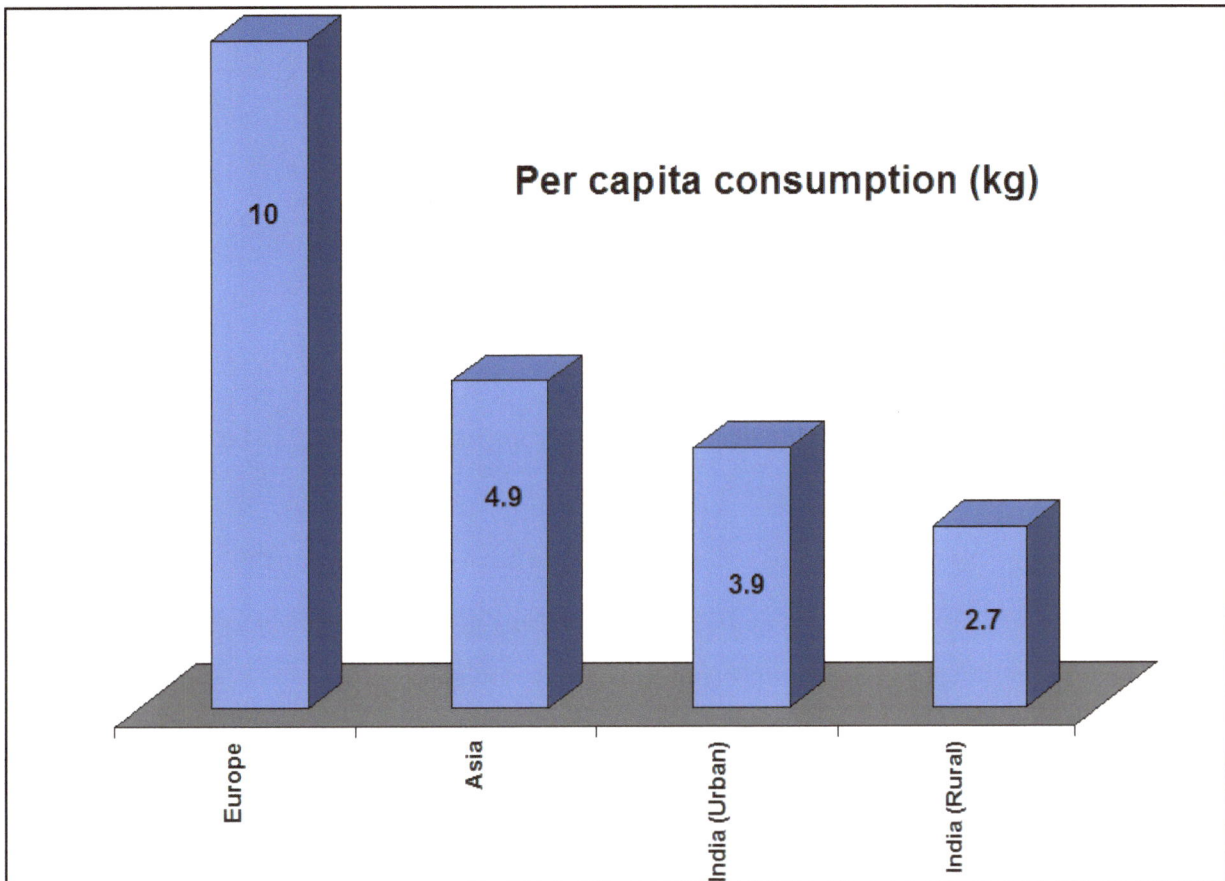

Per capita consumption (kg)

Europe	Asia	India (Urban)	India (Rural)
10	4.9	3.9	2.7

Specific Attributes that Packaging must meet for Detergents

1. Product Related — Packaging should prevent moisture loss / ingress, retail perfume, preclude microbial attack, protect the shape, logo and colour of product and retail overall presentation of the product.

2. Packaging Machine related — Packaging material and design should be such that is matches with the machinability, scuff resistance, heat seal-ability and molding ability

3. Functionality Related — Packaging should provide resistance to external environment, stress crack resistance, ease in dispensing, ease of opening, easy in handling, storage and transportation, sharp on display communications, have a good shelf appeal, favorable on cost / benefit ration.

Different types of packaging are being used according to the price segment and product forms of detergent and also to give visibility on the brand.

Eg. A standard pillow pouch for a discount segment; a pillow pouch with attractive graphics and carry options for a mid range product; and a gusseted stand up laminated pouch for a premium quality

Various packaging formats currently being used-

1. For BARS - Envelope wrapping, Flow wrapping, Flexible pouches. These are primarily poly-coated or lined papers. Also waxed wrappers are used.

2. For LIQUID detergents — Bottles, Refill Packs, Flexible Standy pouches.

3. For POWDER detergents — Cartons (generally poly lined from inside), HDPE Jars, Flexible packs (laminated or monolayer)

4. Outer packaging — This is generally in 3 or 5 ply corrugated boxes depending on the stack load and total weight of carton.

5. HDPE woven sacks — may be used for bulk supply of powdered detergent

Merits and demerits of different packaging materials-

1. Envelope Wrapping — Most preferred type for bars
    a. Gives good presentation and shelf display
    b. Pilfer proof
    c. Amenable to high speed packaging
    d. Larger supplier base
    e. Average moisture and odour barrier is not much
    f. Anti fungal treatment to the substrate is must

2. Flow Wrapping
    a. Wider choice of material substrates
    b. Excellent WVTR and gas barrier properties
    c. Pilfer proof

d. Excellent printability

e. Good display values

f. Machinability may be an issue

3. Carton

a. Premium look

b. Excellent display value

c. Average barrier properties

d. May need inner liner or anti fungal treatments

4. Flexible Packs

a. High barrier properties

b. Wider choice of materials

c. Premium appeal

d. Excellent graphics

e. High shelf appeal

f. Pilfer proof

g. Refill and single dose options available

h. Non reusable

i. Not recyclable unless made of single substrate

5. Bottles

a. Good display value

b. Consumer convenience in terms of storage

c. Pilfer proof

d. Lower costs

e. Reusability and recyclability possible

f. Average shelf display

Need to Innovate - Detergents (as product) and Packaging

To cater increasing demand of modern customers, detergent brands are coming with latest technologies and there is lot of scope in product innovations so brands are innovating their products in terms of easy usage, hassle free washing, unit doses and concentrate forms. Simultaneously packaging requirement will also change for new innovative product forms. Lot of new materials, formats, styles, designs and technology is required to cater the needs of these brands and new-gen products.

To understand these changing trends in new technologies in detergent variants and their packaging we have come up with this innovation report. The book covers packaging innovations in detergent sector that has happened across the globe. Few of the innovations are interesting concepts that may be used for variety of products in detergent category. The book also covers some of the new detergent concepts running in global the markets.

LIQUID DETERGENT

Manufacturer/Designer

GCS Massmould ltd
Tel: +44 1480 217323
info@gcs.com
www.gcs.com/brands/massmould

- This is an innovative closure for packaging liquid detergent.

- The product is packed in unique multifunctional bottle having an easy-to-open flip-top lid and spout.

- The lid is made up of three-piece bicolor dispensing closure.

- The lid is made up of three-piece bicolor dispensing closure.

- The bottle also has an innovative built-in stain removal ball.

- This stain eraser ball is made from TPE (thermoplastic elastomer) and has molded features which are designed to assist with cleaning and measuring the dose needed.

- The ball is filled with the detergent and is then placed directly in the washing machine thus removing stains with friction effects.

Features:

- 3 piece bicolour closure
- Stain eraser ball
- Flip top lid

DETERGENT PACKAGING INNOVATIONS

Manufacturer/Designer

Crescent Manufacturing Inc.
Tel: 716-337-0145
rfrazer@crescentmfg.net
www. crescentmfg.net

- The Bag-In-Box is environmentally superior way to package liquid laundry detergent.

- This kind of packaging has a plastic bag carrying the product.

- The bag also has an easy dispensing knob and further it is covered with corrugated box.

- The cardboard is derived from trees grown in certified sustainable forests, and printed with soy-based inks giving it a sustainable packaging advantage.

- This bag-in-a-box system comes with an easy to press dispensing nozzle which fits into a perforated opening and assists in easy product take out.

- This pack has been named as "cube" and the pack also contains a cup inside for measuring detergent.

- When compared to a 50-ounce bottle (24 wash loads) of 2x laundry detergent, the Cube contains 86% less plastic, and requires 40% less transport fuel per wash load.

Features:
- Easy dispensing through nozzle
- Measuring cup for accurate dosage
- Sustainable packaging

Manufacturer/Designer

DeVries Global
Tel: 212-546-8549
alafleur@devriesglobal.com
www.devriesglobal.com

- The pack has an innovative cap which is branded as Zap! Cap TM.

- This is a unique pretreat cap with scrubbing bristles to provide a deep-down, cleaning action.

- The cap features two textures; bristles for scrubbing and a flatter portion to spread the detergent around.

- This feature is useful for removing tough stains from areas like collars, wrist cuffs, etc.

- The bottle represents the ultra stain release capacity of detergent.

- As a product, Tide Ultra Stain Release is supercharged with specially formulated ingredients to help remove 99% of everyday stains, including greasy food stains.

- The pack provides consumer convenience since additional brush is not required for removing tough stains.

- This kind of packaging is useful for both hand washing and machine wash where in clothes are just rubbed to removed stains and then washed in machine.

Features:

- Textured cap to give scrubbing effect
- Attractive shelf appeal
- Premium look to the product

DETERGENT PACKAGING INNOVATIONS

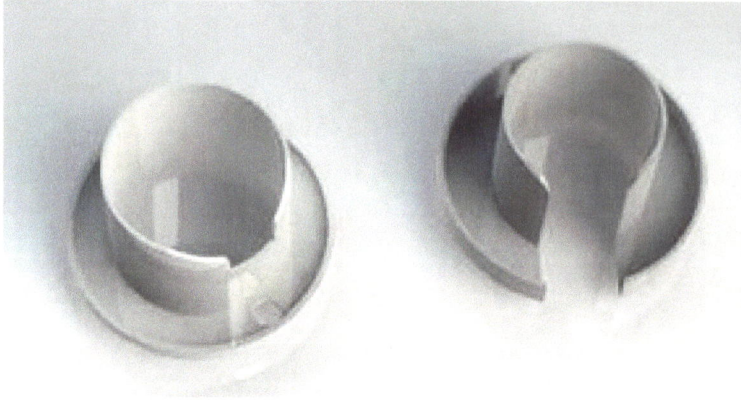

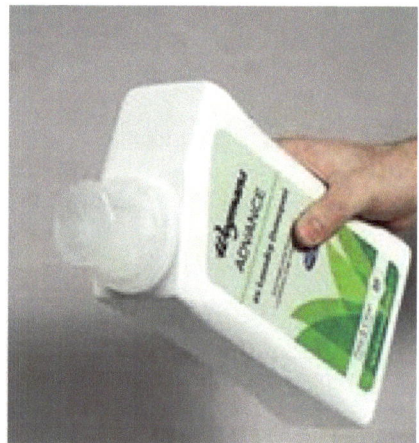

Manufacturer/Designer

Designed by- Katherine Bukys
Sponsored By- Wegmans/American
Packaging Corp.
cias.rit.edu/faculty-staff/154/
student/804

- The innovation is in dispensing mechanism and pack design to save space.

- The pack has a Flip in cap with closure that acts as dispensing cup.

- Bottles are made of EBM HDPE (extrusion blow molded high density polyethylene) and closure from IM PP (injection molded polypropylene).

- Consumer has to just pop the cap and pour the liquid detergent into the transparent lid.

- The design of pack is such that more number of packs may be placed on pallets and hence pallets are better optimized for their space.

- The bottles may be placed horizontally or vertically on pallets to use maximum space.

- User convenience is more while pouring the detergent from the bottle.

- Due to unique design packaging the brand differentiate itself on shelf among other brands.

Features:

- Inbuilt dispensing cup
- Unique shape
- Less storage and transportation space

DETERGENT PACKAGING INNOVATIONS

Manufacturer/Designer

AstraPouch®
Tel: (585) 259-9202
dave@astrapouch.com
astrapouch-na.com

- Flexible packaging is increasing it's share by its preferred usage for various applications.

- This is due to various reasons, one being space advantage.

- Large volume pouches have a significant reduction in carbon footprint and transportation energy over rigid bottles.

- Using this concept, AstraPouch has developed "SuperLam™ lamination" which provides enhanced burst and seal strengths over current laminations.

- The integrated nozzle in the pouch provides ease in dispensing the detergent from a large volume pouch.

- Finger cuts given at the top supports easy handling by the consumer.

- It can be used for wide range of liquid products including oils, condiments, shampoos, detergents, and beverage products.

- Available in sizes of 750ml, 1.0L, 1.5L, 1.75L and 3.0L.

Feature:

- Space advantage
- Easy dispensing
- Easy to carry and handle
- Good option for packaging versatile liquid products

DETERGENT PACKAGING INNOVATIONS

Manufacturer/Designer

Sunbelt paper packaging ltd
Tel: 205.222.3122
chrisrivers@sunbeltpaper-packaging.com
www.sunbeltpaper-packaging.com

- When we talk about liquid detergent, there is always a chance of spilling specially from bulky packs.

- To prevent this, Sunbelt has developed a zero spill packaging for Gain Brand of their liquid detergent.

- The container is made up of HDPE (high density polyethylene) and have two openings for consumer convenience.

- One opening has a knob for controlled detergent flow and avoiding from spilling, while another opening is for uncontrolled liquid flow and re-filling of the container.

- The product is dispensed after pressing red knob.

- Transparent PP (polypropylene) cup may be used for measuring liquid detergent.

- The cup has gradations on different capacity levels.

- Ultra Concentrate Liquid is formulated with Pure Clean Technology™ for whiter and brighter colors on all kind of laundry items.

Features:

- Two side opening for controlled and regular dispensing
- In built measuring cup
- Useful for bulk packaging
- Good choice for re-fill container option

DETERGENT PACKAGING INNOVATIONS

66 LOADS

seventh
GENERATION
™

Natural 4X
Laundry Detergent

Geranium Blossoms & Vanilla

Scent Derived from Whole Essential Oils

50 FL OZ (1.56 QT) 1.47L

66 LOADS

seventh
GENERATION
™

Natural 4X
Laundry Detergent

Free & Clear

Free of Dyes & Fragrances

50 FL OZ (1.56 QT) 1.47L

DETERGENT PACKAGING INNOVATIONS

Manufacturer/Designer

Ecologic Brands
Tel: NA
info@ecologicbrands.com
www.ecologicbrands.com

- This is an interesting packaging for a product like liquid detergent.

- It is a paper bottle with enclosed plastic bag.

- Liquid detergent is in a LDPE low-density polyethylene (LDPE) bag enclosed in bottles made of a board-stock shell.

- Inside bag has a resealable LPDE spout, and the outer cap is of polypropylene.

- The container consists of a molded pulp outer shell made from 70% recycled board and 30% old news papers.

- Using paper for this container results in a 66% reduction in plastic use per bottle, compared with most other 100-oz laundry detergent bottles.

- The bottle may be again with fitted with the refill pouches.

Features:

- Sustainable packaging
- Gives refill option
- Safe and secured packaging for liquids

DETERGENT PACKAGING INNOVATIONS

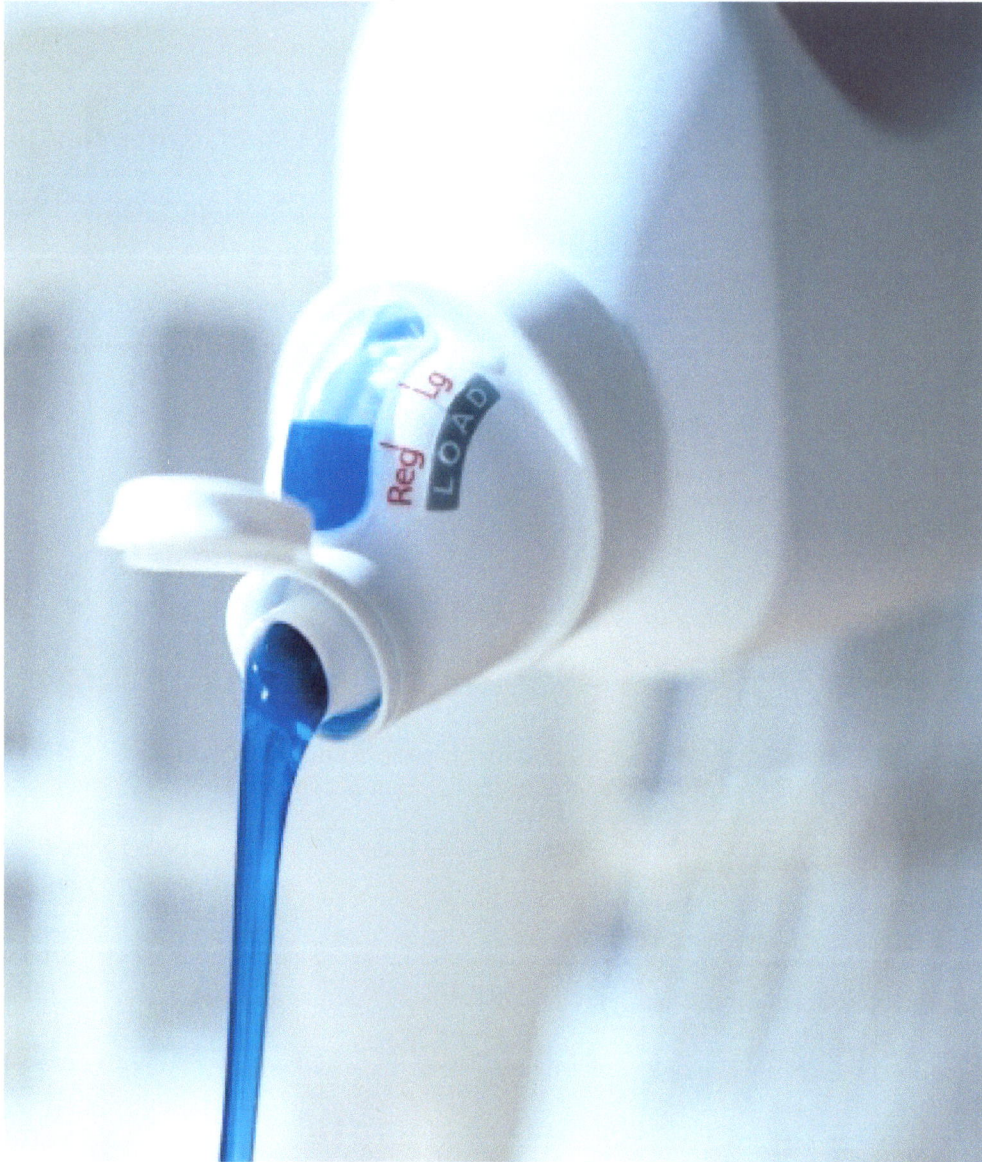

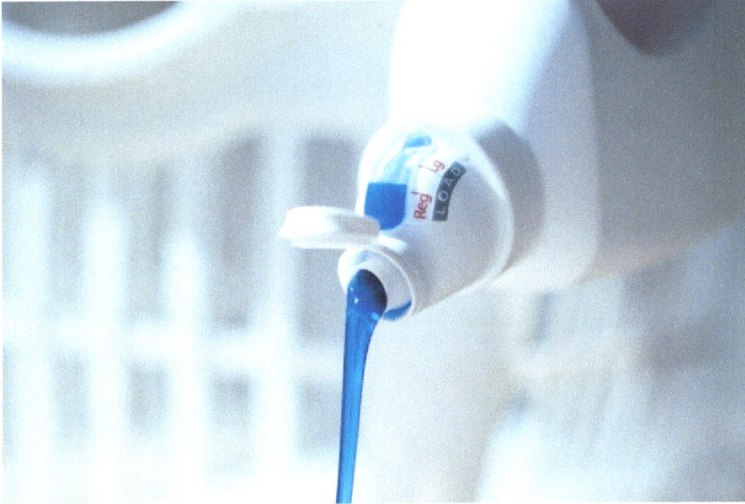

Manufacturer/Designer

Perimeter Brand Packaging
Tel: 508 466-8430
info@perimeterbp.com
www.perimeterbp.com

- All liquid detergents in market are coming in screw cap, snap fit cap or in squeeze form but consumers get problems while pouring the liquid from bottle and also for measuring.

- "Measure as you pour" technology eliminates the need for measuring cup.

- It is designed for freehand dispensing of liquid from a container.

- It shows how much liquid is being poured out by the use of a clear indicator for different loads.

- The consumer can easily dispense the recommended quantity of product while enabling some control to adjust to their preference.

- It can be used for products like edible oils or liquid concentrates, or for pouring household liquids, like detergents or cleaning products.

- Unique design allows for on-shelf differentiation at point of purchase.

Features:

- Freehand dispensing of detergent
- Accurate dosage
- Unique idea for versatile liquid products

DETERGENT PACKAGING INNOVATIONS

DETERGENT PACKAGING INNOVATIONS

Manufacturer/Designer

P&G Panama
Tel: (800) 332-7787
Email: mediateam.im@pg.com
www.pg.com

- The bottle includes a polypropylene dosing cap that can be removed from the body of the pack and an integrated 'visi-strip' which shows consumers how much gel is remaining in the pack.

- Ariel promote washing from water as cold as 15degc which is increasingly important to consumers today to help the planet be environmentally friendly.

- The packaging also reinforces this concept with the rounded shape which links to the shape of the planet.

- The plain white bottle and simple colour schemes suggest it is pure and hasn't got any harsh ingredients or chemicals.

- Brand is targeting this gel for low temperature wash areas.

- As per P&G study, this compact recyclable pack uses up to 45% less packaging than any other Ariel liquid detergents, allowing more bottles per case, pallet and truck to be transported and reducing carbon emissions as fewer trucks are needed.

- The container may be made from moulded high density polyethylene with a polypropylene flip top cap and.

Features:

- Visi-strip for viewing quantity
- Unique shape to give brand identiy

Manufacturer/Designer

The Australian Pouch Company Pty Ltd
Tel: +61 2 8852 2660
info@auspouch.au.com
www.auspouch.com.au

- Australian Pouch Company launched unique bottle shape pouch for liquid detergents.

- This is exactly in the shape of a drinking water bottles pouch.

- The pouch has a spout for opening convenience.

- To add in consumer convenience, a hook is inserted on the top side for hanging purpose.

- Pouch is in specific bottle shape for carrying convenience.

- This pouch gives unique stand out on shelves while among all bottles.

- The pack may be targeted to frequent travelers, hostel-ites that requires low product quantities but handy packaging.

Features:

- Easy in handling
- Easy dispensing
- Useful as a retail pack or a small sku pack

DETERGENT PACKAGING INNOVATIONS

say hello to
method laundry

Manufacturer/Designer

Method ltd.
Tel+44 (0) 207 788 7904
talkclean@methodproducts.co.uk
methodhome.com

- This is a concentrated liquid detergent launched by Method Laundry Detergent.

- This is with a Smartclean Technology which is equivalent to an 8x concentrated formula compared to conventional standards.

- The new detergent features an easy to use pump on the top. For usage it is suggested to pump four times i.e. 4 pumps = 1 load of laundry.

- Method's new detergent is easy to hold in one hand due to its weight.

- It is really easy to use owing to its highly functional pump.

- Overall the pack minimizes material usage due to compact packaging and also reduces waste and annoying detergent drips.

- It is easy to store also due to its small size.

- The packaging uses 36% less plastic than their previous formulation and is made from 50% post-consumer recycled HDPE (high density polyethylene plastic).

Features:

- Compact packaging for concentrated liquid
- Less material usage
- Environment friendly packaging
- Direct and easy dispensing

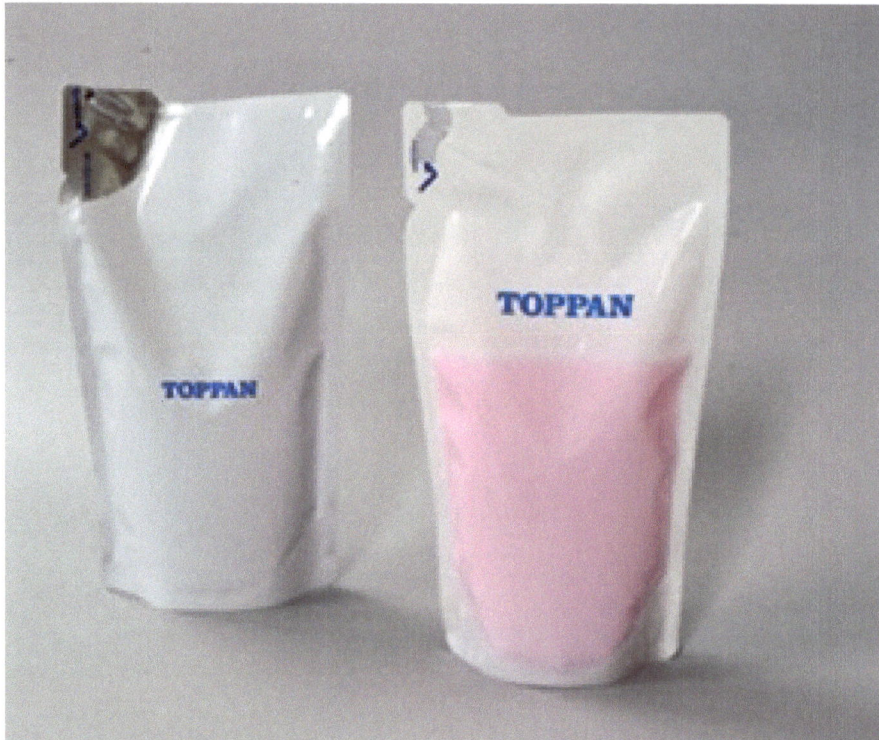

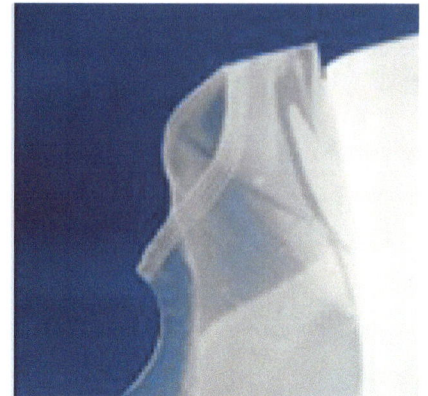

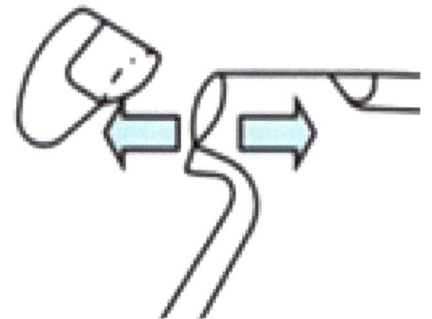

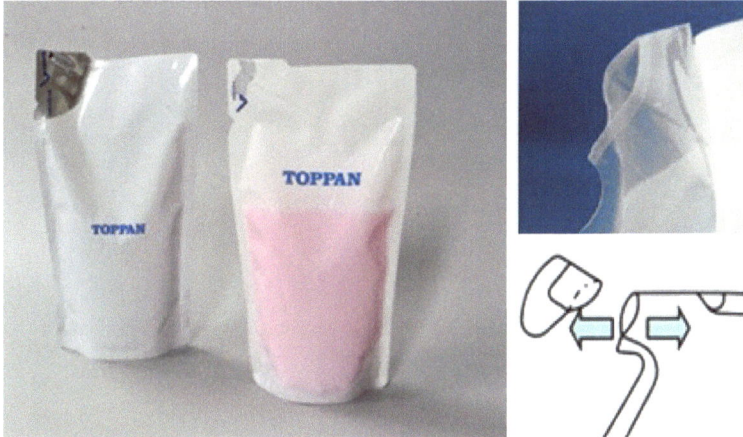

Manufacturer/Designer

Toppan Printing Company Ltd
kouhou@toppan.co.jp
Info.i@toppan.co.jp
www.toppan.co.jp

- The Sosogi Jozu, which means "easy pour", is a stand-up pouch in which the ease of pouring the content has been improved.

- This kind of Stand-up pouch has a structure that includes a welded seal at the top, to form a spout.

- The consumer needs a scissor to cut a part of one of containers before he can start pouring the content.

- This pouch has a pour opening created by folding back the top section of the pouch instead of sealing it.

- By adjusting the shape of the so-formed spout, it has also been possible to improve the ease of opening.

- So giving it an opening and reclosing feature through same spout.

- Due to being a stand up pouch, it gives good shelf presence.

Features:

- Easy in pouring
- Easy opening

Manufacturer/Designer

Aekyung Inc
Tel 8228181700
info@aekyung.inc
www.aekyung.co.kr

- Aekyung, launched a liquid laundary detergent "LiQ" which requires less detergent for a load of laundry as it is in a concentrated form.

- The container cap can be used as a measuring cup and put inside the washer to prevent clothes from getting tangled.

- Consumers were able to use less detergent as they can pump the right amount into the bottle cap and leave it inside during the wash cycle, which also prevents clothes from getting tangled inside the washer.

- It saves energy that is required for packaging, shipping and minimizes waste because just half of what used to be required is needed for a wash.

- Basically in one single packaging, it has all features of packing a concentrated detergent with a measuring and wash- aiding cap.

Features:

- Pumping gives easy dispensing
- Cap itself serve as a wash aid
- Profiled shape give attractive shelf appeal

Manufacturer/Designer

Colgate-Palmolive Pty Ltd Sydney
Tel: +61 2 9229 5600
Email: NA
www.colgate.com.au

- Colgate has developed Smart Shot range to innovate in the new "in-wash" section of the detergent category and branded it as "Dynamo" & "Cold Power".

- It also has an innovative packaging. The innovation is actually in it's dispensing mechanism.

- The bottles contains a flip top cap in a closure that also acts as a dispensing cup.

- Consumer has to just pop the cap and pour the liquid detergent into the transparent lid which is then placed in the machine as such.

- The design of pack is such that it may be placed from both the directions on the shelf.

- Using this advantage brands may print labels in two different directions on front and back panels respectively.

- From the small opening of flip top, it may also be used for spot application on the stained area of clothes.

Features:

- Bidirectional branding
- Unique shelf appeal
- Unique closure to give easy dispensing

DETERGENT PACKAGING INNOVATIONS

DETERGENT PACKAGING INNOVATIONS

Manufacturer/Designer

Ansell Healthcare Europe N.V.
Tel: + 32 (0)2 528 74 00
info@eu.ansell.com
www.ansell.com

- The award-winning work focused on graphic restyling for the global range of branded liquid products.

- Originally this pack is designed for lubricating gel for intimate products but this may be useful for any kind of concentrated or regular kind of liquid product like detergents.

- The pack has special packaging restyling for POP (point of purchase) appeal.

- The pack has been studied in detail to distinguish, intrigue, convey reliability of the product being packed.

- At the point of sale, the elegant silhouette catches the eye.

- Easy pumping mechanism gives ease in dispensing.

- The top design has a circular loop through which the bottle may be hung on the wall as well.

- This packaging idea may be useful for fabric conditioner and fabric deodorants.

- The bottle may be made of HDPE (high density polyethylene) with a cap made of HDPE of PP (polypropylene).

Features:

- Useful for fabric care category
- Unique shape
- Being compact occupies less space

Manufacturer/Designer

P&G Panama
Tel: (800) 332-7787
Email: mediateam.im@pg.com
www.pg.com

- This is an interesting idea to introduce a different product from the same brand.

- In introducing the cross-promotional package for Tide and Febreze Freshness, Procter & Gamble opted for a decorated cap with a shrink-sleeve label.

- This innovation is intended to attract consumer's attention and to help them easily distinguish Tide with Febreze Freshness from other products in the Tide lineup.

- Label graphics are gravure-printed in five colors.

- Color-coded caps indicate the scent varieties. Matching colors and an icon appear on the bottle label to reinforce the brand's identity and indicate the product scent.

- A decorated cap gives the laundry detergent package distinction in a category in which the bottle caps are typically undecorated.

- This Packaging innovation may help user companies in promoting their support brands with the main product.

- Different colour combination on the caps help identifying variety of product variants available.

Features:

- Wonderful idea for cross product promotion
- Decorated cap gives quick differentiation on shelf

Manufacturer/Designer

Earth Friendly Products Proline Ltd
Tel: 1-800-592-1900
LUKE@ECOS.COM
www.efpproline.com

- A company 'Earth Friendly Product Proline' has launched a laundry detergent in a bulk container of 5 gallon.

- The pack has an easy opening cap to dispense liquid detergent from big gallon container.

- The container is made of HDPE (high density polyethylene).

- The cap is also made of HDPE. Closure has a two side opening for dispensing and filling through twist nozzle and the screw closure respectively.

- The detergent may be dispensed from the knob and refilling may be done by unscrewing the round closure from the container.

- Also since the pack is such it takes more of a horizontal space than vertical, it may save space when placed in pallets.

- Stacking on one container over the other is also possible.

Features:

- Two way opening for filling and dispensing
- Save space during transportation

DETERGENT PACKAGING INNOVATIONS

Manufacturer/Designer

Contact information not available
www.pg.com/pt_BR

- Procter & Gamble in Brazil has released a novel 3L bottle of Ariel Automatic Liquid Detergent.

- The innovation in this bottle is it's handle. Generally the carry handle ia a part of the closure which seldom breaks due to handling pressure.

- In this innovation, handle is an integral part of the container neck itself. Thereby, providing moldings for better grip.

- Generally handle is the part of closure but here it is part of bottle for proper holding grip and strength.

- The handle bottle has fast manufacturing rates.

- The bottle is made of PET (polyethylene terephthalate or commonly known as polyester).

- Molding a handle in the neck gives it adequate strength to handle heavy weight product.

Features:

- Secured handling of bulky container
- Easy grip

Manufacturer/Designer

TechniPac,
Tel: 507.665.6658
gmelchoir@technipacinc.com
www.technipac.com

- Technipac has developed "PresSURE-Lok", which is an ideal solution for controlled dispensing of liquid products in the form of shampoos, personal care products, detergents, energy products, and even beverages and condiments.

- Controlled-Dispensing Technology combines a flexible pouch with a flexible and self-contained fitment, eliminating the need for a separate dispensing component.

- This actually is a flexible laminate with an intermittent seal that opens in a dispensing closure like pocket.

- The user simply squeezes the flexible pouch to open the seal and this allows liquid product to flow into the dispensing chamber. The package then automatically reseals itself for future use.

- The idea is similar to the dental wash wherein a rigid bottle is there that allows fixed dosage liquid to flow in the dispensing cup. The only different is this is a flexible pouch.

- The overall concept is giving an option of material saving in terms of plastic being used.

Features:

- Controlled dosing
- Flexible pouch gives material saving
- Light weight packaging

DETERGENT PACKAGING INNOVATIONS

Manufacturer/Designer

LG household and Healthcare
Tel: 81-3-5537-7311
Email: NA
www.lgcare.com

- This is a shaped pack designed specifically to serve a purpose.

- To illustrate that the detergent can be safely used on baby clothes, these containers are designed in the shape of a standing baby.

- The packaging design conveys immediate message and indicates which product is used for innerwear and which for outerwear.

- Also just by looking at this product, consumer immediately relates that the same is in "baby care category".

- The pack ha s a nozzle in the lid that allows users to squeeze the container while pouring the detergent.

- The bottle is made of HDPE (high density polyethylene) with HDPE cap.

- The lid also allows user to measure the detergent amount.

- Unique packaging shape gives brand differentiation when placed on shelf as "detergent category".

- This pack has won RED DOT21 Design award.

Features:

- Unique shapes that relates to the product packed
- Attractive design and shelf appeal

128 Loads

Simply Soap Berry™
10x Power

Nature's Own Powerful
LAUNDRY DETERGENT

✓ Deep Clean   ✓ Ultra Concentrated
✓ Natural   ✓ Biodegradable
✓ Softens Naturally ✓ Hypoallergenic
✓ Non-GMO   ✓ Low Cost per Load

Free & Clear

32 fl oz (1 QT) 946mL

128 Gentle Loads

Simply Soap Berry™ BABY
10x Power

Nature's Gentle
LAUNDRY DETERGENT

✓ Deep Clean   ✓ Low Cost per Load
✓ Natural   ✓ Biodegradable
✓ Softens Naturally ✓ Hypoallergenic
✓ Non-GMO   ✓ Ultra Concentrated
✓ Enzyme-Free

Free & Clear

32 fl oz (1 QT) 946mL

DETERGENT PACKAGING INNOVATIONS

Manufacturer/Designer

Simply Soap Berry TM
Tel: 888-229-2816
info@simplysoapberry.com
www.simplysoapberry.com

- Detergent packs either comes with a dosing pump or with closures and spouts.

- These features provide easy dispensing but sometimes user demands innovation.

- To cater the needs of consumer and ease in dispensing, the brand "simply soap berry" launched its detergent in a drop shaped container with a dispensing pump.

- The functionality is similar to shampoo packs. Users simply has to press the pump to allow detergent to flow through the nozzle.

- In the pack labels, its is mentioned to use 2-3 pumps for small loads, 4-5 pumps for medium loads and 6-7 pumps for large loads.

- Since this is concentrated detergent, it is mentioned to use the pumped amount in water, mix and then add to machine.

- Instead of messy detergent cap that requires precision filling, this product uses hand pump which is found to be very easy by consumers. No dosing cups or caps required.

- Drop shaped container is made of HDPE (high density polyethylene) with HDPE or PP (polypropylene) caps.

- The label is an in-mold-label (IML)

Features:

- No requirement of dosing cups
- Simple and effective dispensing
- Unique shape

DETERGENT PACKAGING INNOVATIONS

POWDER DETERGENT

Manufacturer/Designer

Contact information not available

- This is actually a 'concept' for packaging detergents.

- The material is entirely different from the conventional ones.

- Either HDPE containers or laminated flexible packs are currently being used to pack detergents.

- However this concept uses a different base material. Non-woven bags (made of PP) or woven bags (made of HDEP) are being used to pack powdered detergents.

- These can be manufactured in multicolor and white color for product specific requirements.

- Just by looking at the pack, product may be differentiated — intended for colored clothes and white clothes respectively.

- Finger cuts given at the top assist in easy holding of the pack.

- The sizes may vary from small to bulk packs.

- Multicolor printing options are available on these bags.

- Due to single raw material, the packs are easily recyclable.

- The pack may be used directly as primary pack with the liner inside or may contain a separate poly-bag containing detergent (depends on choice of the manufacturer).

Features:

- Recyclable option
- Colour choice gives product differentiation on shelf

Manufacturer/Designer

P&G Panama
Tel: (800) 332-7787
Email: mediateam.im@pg.com
www.pg.com

- Flexible packaging is heavily being used for packaging of powdered detergents.

- Companies regularly look for options to make the packs recyclable or reduce material usage and make the pack environment friendly.

- With the same objective 'P&G –Panama' met a need of the Mexican market.

- The pack is designed with PE (Polyethylene) and gave it a high gloss using OPP (oriented polypropylene).

- This two ply laminate makes the pack easy to open and high gloss owing to OPP properties.

- And by this the structure was downgauged by 21 percent, eliminating 111 tons of material/year and saving 20 tons of CO_2 that is directly attributable to the films' extrusion process.

- This may typically replace a premium brand that uses a three ply laminate or a two ply laminate using PET (polyethylene terephthalate) and PE.

Features:

- Easy tearing for frustration free opening
- Downgauging

DETERGENT PACKAGING INNOVATIONS

Manufacturer/Designer

Aekyung Inc
Tel 8228181700
info@aekyung.inc
www.aekyung.co.kr

- Prize-winning laundry detergent pack, designed to resemble with a washing machine.
- The idea is to develop a packaging that matches the household appliance in which the product is to be used.
- The product is made of HDPE (high density polyethylene) with HDPE cap.
- The pack has been given a transparent window on front panel that again resembles a front load washing machine design.
- Also from this circular window of plastic pack, consumers can see the product remaining.
- A die-cut label has been provided on front panel that looks like a control panel of the washing machine.
- To assist in scooping, a space has been given on lid where spoon has been fixed.
- When the pack is kept on the shelf, it gives a look of a washing machine filled with clothes.
- Also through the transparent window it is easy to see the quality of product.
- Sometimes, brands add "coloured beads" to assist in tough stain cleaning or some other special features.
- Through this window, it is easy to highlight the product USP to the consumers without even doing a word-of-mouth marketing for the same.
- It is like packaging speaks itself for the product packed.

Features:

- Very creative design
- Immediately relates with the product usage
- See through window serves multiple functions

DETERGENT PACKAGING INNOVATIONS

DETERGENT PACKAGING INNOVATIONS

Manufacturer/Designer

Seal-Spout Corporation
Tel: (908) 647-1900
info@sealspout.com
www.sealspout.com

- This innovation is an extension on the existing cardboard packaging for granular or powdered products.

- High-quality pour spout has been added on paperboard and corrugated box.

- Spout protect powdered detergent from contamination, moisture and accidental spills. It allows convenient dispensing of the powdered product.

- Pour spout for is lightweight and durable. Seal-Spout are manufactured in a wide range of colors, sizes, shapes and materials and can be customized to meet customers need.

- Spouts costing about $.01 per carton, depending on the application.

- Spout can be used not only for detergents but for many other powdered products.

- By the means of this spout, it is possible to make a carton packaging reclosable and usable for repeated opening and closing.

Features:

- Pour spout gives easy dispensing
- Prevents contamination of product
- Gives repeated opening and closing feature

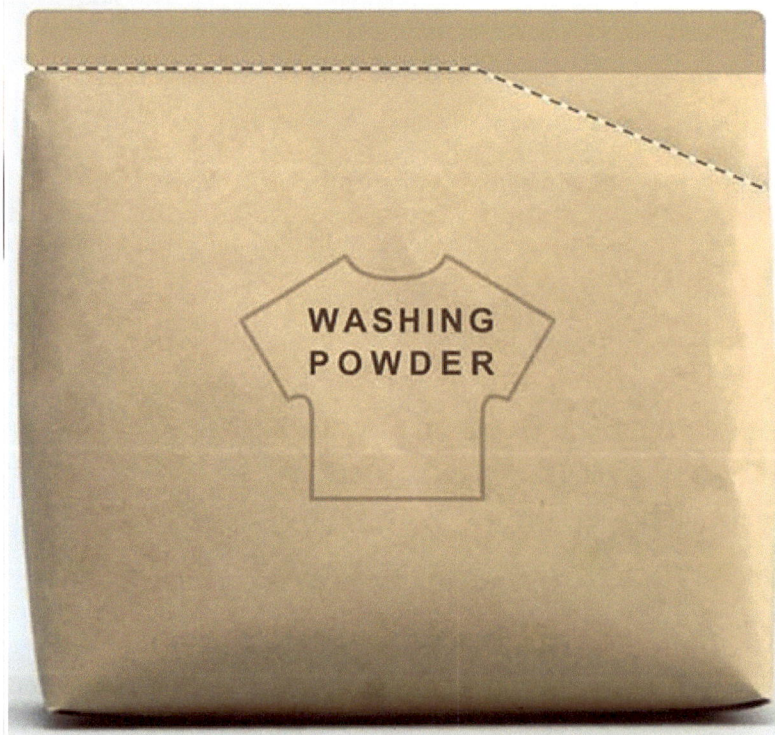

WASHING POWDER

DETERGENT PACKAGING INNOVATIONS

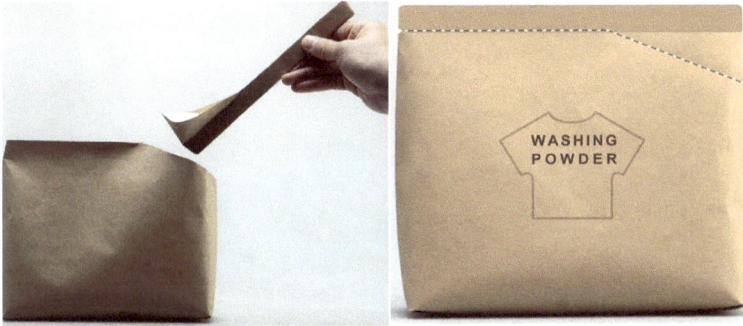

Manufacturer/Designer

Designed by- Yang Guo, Qiaoge Yang & Wenju Wu
Tel: +86-510-85913623
yaoxin@jiangnan.edu.cn
english.jiangnan.edu.cn

- This particular design is a concept designed by a designer team for a pack of powdered detergent.

- The pack is designed with a paper substrate giving sustainable feature to the packaging.

- Also the design is in such a way that there is no need of additional spoon for scooping detergent.

- Instead once the pack is opened, the tear-off part itself becomes a scoop.

- The top of each package is perforated to tear into a handle, with one corner of the bag, which customer can tear off for pouring.

- Detergent boxes currently come shipped with a little plastic scoop, which is an added cost for the manufacturing company.

- Also placement of spoon in each bag involves an additional operation on filling / packaging machines.

- Clean front panel of the pack gives enough space for branding and labeling.

- This may be used as refill pack as well.

Features:

- Environment friendly packaging
- Light weight
- Material saving

Manufacturer/Designer

Designed by- Steffen Westbye

Tel: +47 922 09 790

steffen@westbyedesign.com

www.westbyedesign.com

- Gable top cartons are in much use these days for liquid packaging of product like milk, juice and even soups.

- The concept is now being ideated for detergent category and this is a concept design by a designer.

- The designer revives the old P&Gs brand heritage with being inspired by the product design of the Victorian era, and the graphic style of early 1900s.

- By creating a functional, familiar and easy-to-handle packaging, the design keeps consumers life a little better.

- Easy four side foldable pack which gives ease in manufacturing and printing.

- Due to more fold ability, more number of packs can be loaded in the same space thus reducing transport and storage cost.

- Aesthetics of the pack keeps it stand out on shelves on supermarket.

- Being in a paper and paper board substrate the look is more eco friendly compared to the plastic packaging.

- No doubt this packs gives differentiated look in the shelf.

- Unlike other products where multi layered structures are used to give aseptic packaging, this pack may be designed simply by a paperboard.

- Only similarity with the gable top carton is it's structural look

Features:

- Sustainable packaging
- Unique shelf appea
- Easy handling
- Saves space during transportation and storage

DETERGENT PACKAGING INNOVATIONS

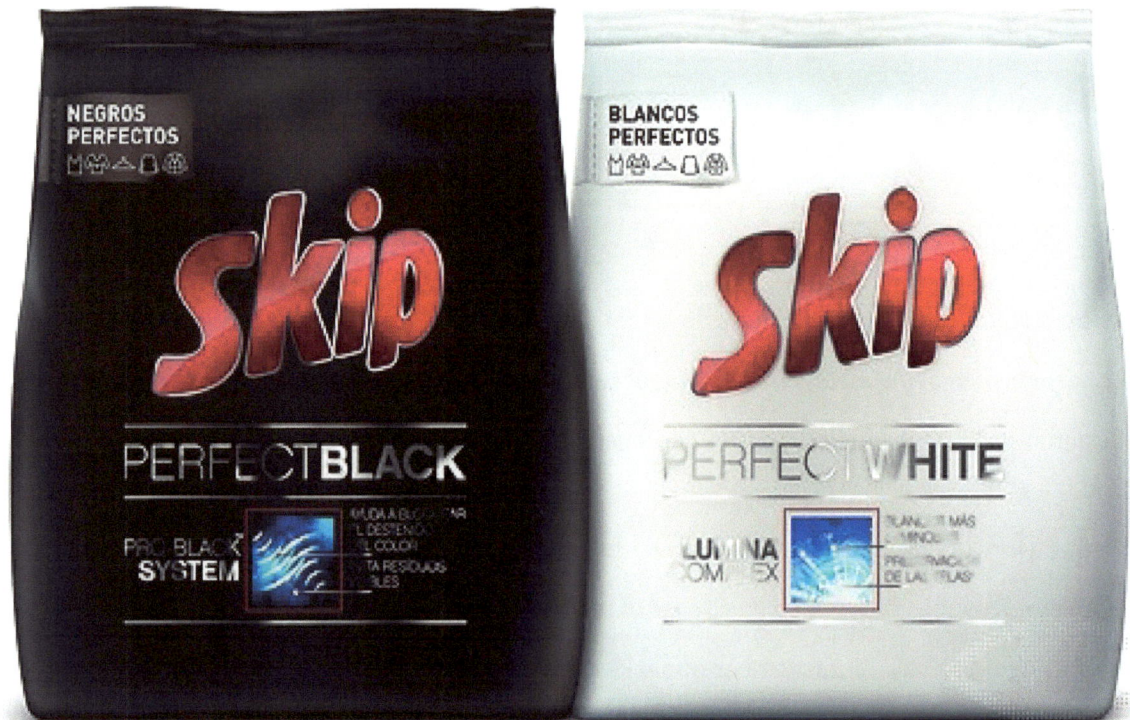

DETERGENT PACKAGING INNOVATIONS

Manufacturer/Designer

REX Design
Tel: NA
Email: NA
www.dexigner.com/news/18365

- This is mainly a design innovation which clearly resembles with the type of usage of the product.

- Packaging for the new range of Skip powder detergent is in two formats - Perfect Black / Perfect White.

- This directly relates to the end product i.e. used for black and white clothes respectively.

- Very elegant typography and "clean" icons that reveal the technology behind the product.

- A plus point is the packaging effect caused by the chosen substrates.

- The pack has been given a matte finish with glossy silver text.

- Overall reinforcing the product premium positioning.

- This innovation gives an idea that by simply changing the printing and giving specific effects, brand may be given a completely changed and differentiated premium look.

Features:

- Attractive printing effects
- Premium look to the product

Manufacturer/Designer

Ampac Flexibles GmbH
Tel:+41 34 448 44 45
adrian.haeberli@ampaconline.com
www.ampaconline.com

- The recyclable, recloseable stand-up pouch application developed in cooperation with biodegradable laundry detergent marketer "Savvy Green™, St. Petersburg, FL".

- It has won an FPA Gold Award in Environmental & Sustainability Achievement for Ampac Flexibles.

- The pouches are made from a proprietary coextruded HDPE (high-density polyethylene) blend and a small amount of non-HDPE resin to enhance pouch material processing.

- Included in the improvements in this product are the stiffness, toughness for drop and distribution, seal properties for improved packaging efficiencies, and moisture barrier for dry products.

- Pouches can be recycled along with other HDPE products in municipalities with recycling programs as well as at retail stores with drop-off stations for HDPE shopping bags.

- It also includes high definition flexographic printing for added shelf "pop"!

Features:

- Biodegradable packaging
- Enhanced functional properties

DETERGENT PACKAGING INNOVATIONS

Manufacturer/Designer

THE LAUNDRY AT LINENS LIMITED, INC.
Tel: 414-223-1123
lizb@linenlaundry.com
www.linenlaundry.com

- This gives an innovation in terms if break through technology wherein packaging material has been totally shifted from plastic (rigid and flexible) and paperboard to metal.

- 'The Laundry at Linens Limited' launched laundry powder in Tin container.

- A normal zinc coated tin can with snap fit cap.

- This packaging gives differentiated look on shelf since rarely we see any detergent being packed in tin cans.

- Powder is packed in polythene bag which is thermally sealed and placed inside the container.

- A plastic spoon is also there for measuring purpose.

- Adhesive paper label is used for decoration purpose.

- This container may be refilled by refill pouches available in market.

- The container may also be reused for domestic purpose.

- Shiny / glossy metal surface enhance premium look of the detergent.

Features:

- Material used is entirely different from conventional ones
- Refill option possible
- Premium look to the product

DETERGENT PACKAGING INNOVATIONS

Unilever de Argentina S.A.
Tel: 0800-888-6666
Sales@unilever.sa
www.unilever.com.ar

- The pack has been developed by Unilever for its brand "OMO".

- This contains a micro fluted carton with high quality printing.

- Spot UV coating has been done on characters of brand name for extra effects.

- Aqueous coating on surface has been given to provide scratch resistance and aesthetic look.

- Detergent powder is into hermetically sealed plastic pouch which is placed inside the box. So this is a kind of bag-in-box packaging for powdered detergent.

- Flexible pouch gives adequate barrier against moisture.

- Outer carton gives a good substrate for branding and decoration giving high shelf appeal.

Features:

- Attractive graphics and print effects
- Bag in box style of packaging

DETERGENT PACKAGING INNOVATIONS

Manufacturer/Designer

Studio Armadillo is located in Tel-aviv
Tel: 972-03-5298150
contact@studioarmadillo.com
www.studioarmadillo.com

- This innovation drives re-usage functionality of a detergent pack.

- The pack is designed in a shape of a triangular box that contains detergent powder.

- When the detergent is finished the box can be reused as flower pot and in this way the packaging has a second life.

- The shape of the pot is modular, so different boxes can form a composition of flower.

- The packaging is attractive and colorful – it adds an additional value of sustainability and fun to this series of products.

- When different product variants are placed on a shelf, there is a possibility to arrange them in such a way that the packaging itself gives a look of a POP dispenser.

Features:

- Unique shape
- Reusability of packaging
- No / zero packaging waste

DETERGENT PACKAGING INNOVATIONS

Manufacturer/Designer

BZPI ltd
Tel: +375 177 75 26 96, 74 48 09
bzpi@bzpi.by
www.bzpi.by

- This is again a rigid packaging idea for detergent powder wherein packaging looks like a paint bucket.

- BONUS detergent has been launched in a plastic container with capacity of 10 kg.

- The container is made of HDPE (high density polyethylene) with an IML (in mold label).

- The cover of the container can be closed hermetically.

- Lid has a special tear off option which gives ant-counterfeit feature.

- A handle is fixed from both the sides (just like a paint bucket) to provide handling convenience.

- Large surface gives lot of scope for branding and decoration.

- The packaging is targeted for bulk powder detergent.

- A plastic Spoon is given inside for scooping out the detergent.

Features:

- Anticounterfeit option
- Large surface area for printing and decoration

Manufacturer/Designer

Contact information not available
Lisa Skowyra

- This is a concept pack designed a designer for powdered detergent.

- The innovation is mainly in terms of 'design'.

- Look of the pack is somewhat similar to the front load washing machine.

- A plastic film has been given in the center to form a window through which product may be seen.

- Front panel may be printed or colored based on the product it contains. Ex; white color front may reflect that the product contained is specifically for white colored clothes.

- Detergent is packed into hermetically sealed poly bag and is placed inside the carton.

- Packaging itself gives differentiation on shelf.

- Transparent window on the carton may be used to display special ingredient like "blue stain release balls".

- Also the graphics are such that it showcase environment friendly characteristics.

Features:

- Attractive shelf appeal
- Bag-In-box format
- Clear window to see product

UNIT DOSAGE

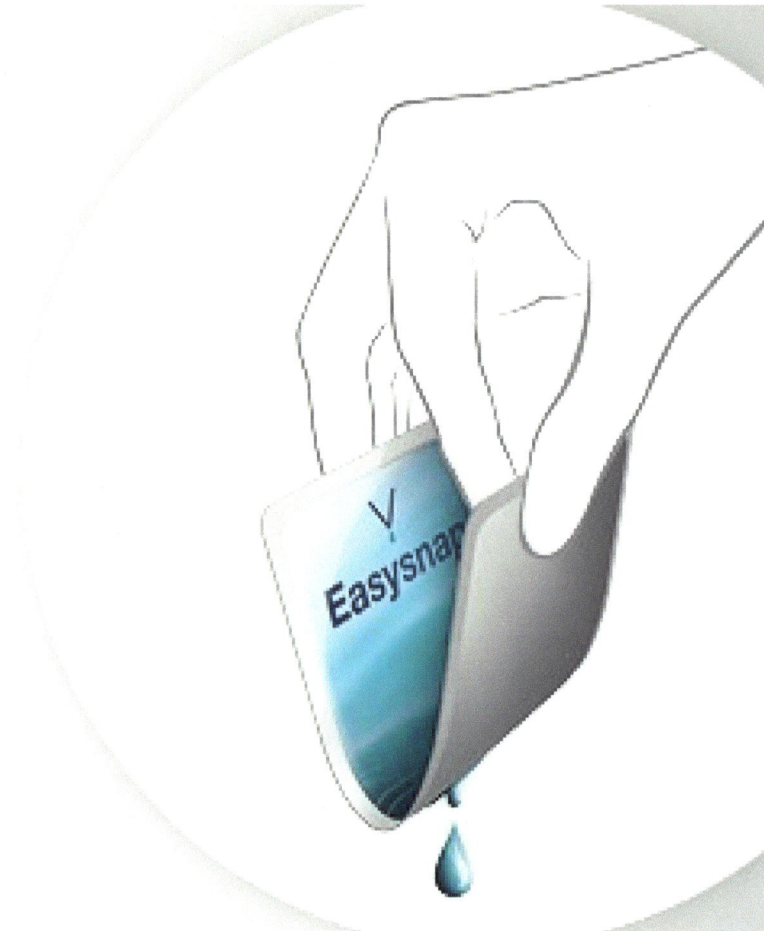

DETERGENT PACKAGING INNOVATIONS

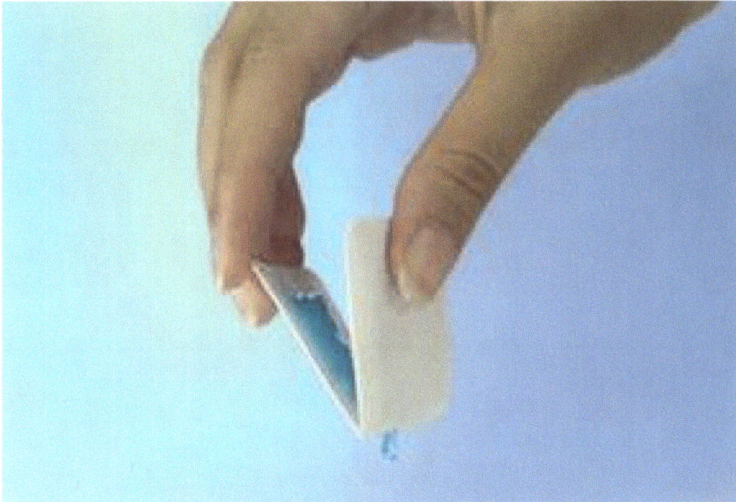

Manufacturer/Designer

Easysnap Technology S.r.l. Italy
Tel: 02325901201
info@easysnap.com
www.easysnap.com

- Easysnap is an innovative, patented and revolutional monodose packaging concept (portion pack from 1 to 25 ml).

- It can replace any conventional sachet in food industry, pharmaceuticals, cosmetic and chemicals.

- Easysnap is an absolute novelty which can replace all the 3 or 4-side seal monodose flexible sachets in the market.

- It is often difficult to open a sachet by wet hands (during bathing or while washing clothes), this innovative technology gives solution.

- With its one hand opening system, Easysnap is suitable to open and dispense in a clean way any liquid product.

- It is perfect solution for liquid detergents, fabric deodorant.

- Mainly used for unit doses like to be used as travel packs

Features:

- Unique dispensing style
- Good option as a on-the-go pack
- One hand opening of sachet

DETERGENT PACKAGING INNOVATIONS

Manufacturer/Designer

MonoSol, LLC
Tel: 219-762-3165
info@monosol.com
www.monosol.com

- These are unit doses in form of "pods".

- The pre-measured unit dose pack uses a specially-developed film that dissolves completely during the the wash, even in cold water.

- This enables consumers to reduce energy used by washing more loads in cold water.

- The unit dose detergent features three chambers especially designed to brighten, fight stains and clean.

- Pod will also have feature like child-resistant closure; to deter children from eating the brightly colored packets that looks like candy.

- These pods may be packed in rigid plastic tub which is made of 25% recycled PET.

- Alternatively can be packed in stand-up bags.

- This reduce plastics use by 50% per load and total packaging material use by 11% respectively.

- This form of detergent may be really convenient for hassle free washing, without any chance of spillage of liquid or powder detergent.

Features:

- Hassle free washing
- Consumer convenience
- No / zero packaging waste

DETERGENT PACKAGING INNOVATIONS

Manufacturer/Designer

The Moderns
Tel: 212.387.8852
themoderns@themoderns.com
www.modernsnyc.tumblr.com

- This innovation is in sustainable category.

- The brand "Berry+" is in a packaging which is 100% plant-based.

- The packs is for laundry detergent packaging for 10 loads.

- Entire packaging is designed by 'The Moderns'.

- Complete detergent and conditioner is filled in small rubber tubes. Each tube is for single unit detergent.

- Outer box is made up of molded paper pulp.

- Entire appearance of the package look like first Aid kit that resembles with the idea of care about laundry.

- Left panel of the box may be effectively used for branding.

- This can be a good option for on-the-go detergent packaging.

- Since the materials used are environment friendly, the pack has a high rating in terms of eco friendly features

Features:

- Biodegradable packaging
- Completely different look

PODs are torn off and used one-by-one until there is no packaging left.

DISAPPEARINGPACKAGE.COM

Manufacturer/Designer

MonoSol, LLC
Tel: 219-762-3165
info@monosol.com
www.monosol.com

- MonoSol has developed a 100% dissoluble packaging for Tide pods.

- The package itself is a sheet of laundry pods stitched together to give a sheet kind of format.

- The packs are printed using soap-soluble ink.

- The POD plastic and the pods both are water-soluble.

- Consumers tear off each POD and use one-by-one. With the last POD, the package itself is gone.

- PODs, instead of being stored loose, are stitched together into a perforated sheet.

- Product details and brand information are then printed directly onto this sheet.

Features:

- No / zero packaging waste
- Innovative way of collating the unit packs together

Manufacturer/Designer

LPK United Kingdom
Tel: +44.(0)20.7317.0800
amy.steinmetz@lpk.com
www.lpk.com

- The packaging has been designed by LPK for P&G's brand 'Cascade'.

- The idea has been inspired from beauty care products by leveraging a "jewelry box" design to make the Cascade detergent a more exclusive brand.

- The container is made of clear PP (polypropylene) that offers a different look compared to the competitive brands being packed in paperboard containers and opaque bottles.

- The pack gives brand differentiation when placed on shelves.

- This product is originally launched for dishwasher pods but can be replicated for detergent also.

- This soft, feminine shape gives women confidence that the product will get their dishes clean in one cycle through the dishwasher, and provides the package with the visual impact that can make it a decorative item on kitchen.

- Pods are made of water soluble film and container has an IML label.

Features:

- Premium look by a secondary packaging
- Reusability as storage container

TRAVEL
SINK PACKETS
PAQUETS DE
VOYAGE POUR ÉVIER

3 USES
UTILISATIONS

HAND WASH IN SINK / LAVAGE À LA MAIN DANS L'ÉVIER

Caution: Eye Irritant. Harmful
If Swallowed. See Caution On
Back Label. / Attention : Irrite
les yeux. Nocif si ingéré. Voir
la mention Attention au dos.

LIQUID DETERGENT
DÉTERGENT LIQUIDE
3 x 5 mL (0.51 FL OZ)

DETERGENT PACKAGING INNOVATIONS

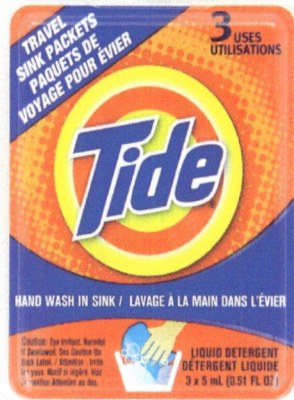

Manufacturer/Designer

P&G Panama
Tel: (800) 332-7787
Email: mediateam.im@pg.com
www.pg.com

- This is an innovative format of unit dosage for detergents.

- The detergent is in form of sachets which is then packed together.

- P&G Launched special sink packaging for detergent sachets specially to carry during travelling.

- During travelling it is difficult to carry powder bag or liquid bottle.

- Just enough liquid detergent in single sachet to wash a few items in the sink itself.

- The combo of three sachet is further packed in a clamshell blister pack.

- The material may be PVC or poly-olefins.

- Clamshell pack has euro hole that gives pack additional feature of hanging on the wall.

- The transparent clamshell provides excellent branding.

- Either individual sachets may be printed else plain sachets may be packed in a clamshell and the card insert may serve the purpose of branding.

- As an alternative to clamshell, carded blister packaging is also an option for this.

Features:

- Easy to carry pack during travelling
- Lot of space for branding

10ˣ **Ultra**

Lavender

32 WASHES

Scratch & Sniff Here

Dizolve®
Laundry Detergent Sheets

SUPERIOR STAIN REMOVAL

PF (he)

DETERGENT PACKAGING INNOVATIONS

Manufacturer/Designer

Dizolve Laundry detergent sheets ltd
Tel: 1800 065 326
info@pascoes.com.au
mydizolve.com.au

- 'Dizolve' brand of laundry detergent has a unique product i.e. detergent in form of sheets.

- These are for single time use and the packaging is simply in a zipper pouch with top seal.

- Zipper allows repeated opening and closing of the pack.

- The pouch is also provided with a Euro hole for hanging on the wall.

- The product sheets are such that they gets degraded (even in open environment) in a span of around 14 days and the product is also phosphate free.

- Packaging is also environmentally friendly since a small and lighter packaging is only required to pack these sheets compared to a heavy and big container for other product variants.

- It is also easy to carry during travelling.

Features:

- Smaller and lighter packaging
- Multi usage through zipper

For your travelling laundry

Step 1

Step 2

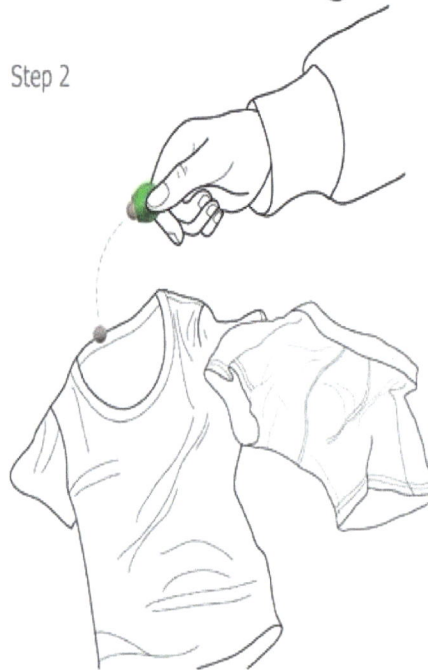

Waterproof paper(biodegradable)

soap pod powder "ball"

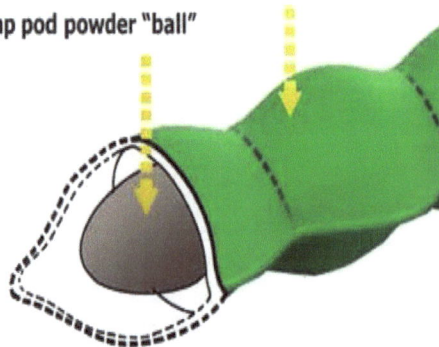

DETERGENT PACKAGING INNOVATIONS

For your travelling laundry

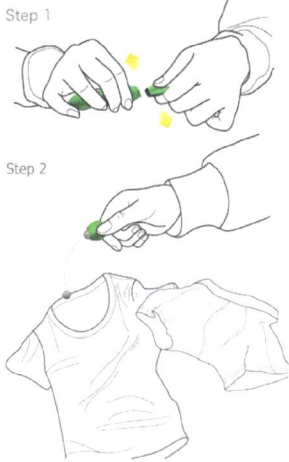

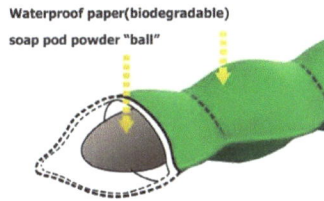

Step 1

Step 2

Waterproof paper(biodegradable)

soap pod powder "ball"

Manufacturer/Designer

Designed by Yingying Zhou, Shijiao Li, and Sicheng Wang
Tel: +86 13916634808
design.zyy@gmail.com

- This product is unique in the category of unit doses.

- The pod has been designed with the objective to solve the detergent carrying issues during traveling.

- The concept of Easy Pod Washer has been inspired by a Chinese traditional honey locust and is called as "soap pod".

- The "soap pod" can be broken down microbiology quickly and causes no harm to the environment.

- The soap pod is grind into fine powder and then shaped into individual balls, just like beans and is then put inside the cover of pod.

- During time of laundry, individual pod may be detached from the rest, the drop is squeezed to remove the ball which is then dropped into the water.

Features:

- Innovative product in unit dose category
- Biodegradable
- Convenient as on-the-go detergent packs

INTERESTING CONCEPTS

DETERGENT PACKAGING INNOVATIONS

Manufacturer/Designer

SmartKlean
Tel: NA
info@smartklean.com
smartklean.com

- The SmartKlean Laundry Ball uses an innovative technology designed to clean fabrics through a physical process instead.

- The laundry ball is filled with four types of mineral-derived ceramic beads and two magnets, each performing different cleaning functions.

- When these components come in contact with water, they form 'oxygenated' water with an increased pH level and an ability to eliminate germs and bacteria.

- Component of ball are Far Infrared Ball, Alkali Ball, Anti-microbial Ball, Chlorine-removal Ball, Magnets and Plastic Enclosure.

- The plastic exterior of the laundry ball is composed of environmentally-friendly thermoplastic elastomer free of BPA and PVC.

- Its shape and weight is designed to create agitation in order to beat dirt and grime off clothing during the wash cycle.

- This whole system in form of a ball is packed in a micro-fluted carton.

- The product is with unique technology and packaging itself form the part of the product.

- Outer carton may be printed with various effects to highlight the usage of product and key features.

Features:

- Differentiated product technology
- Hassle free washing

DETERGENT PACKAGING INNOVATIONS

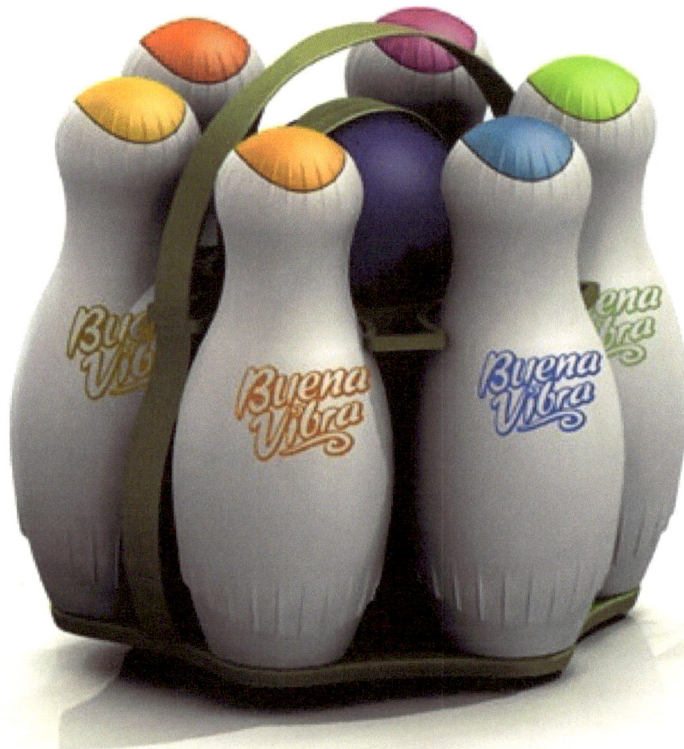

DETERGENT PACKAGING INNOVATIONS

Bow-Bow Detergents six pack

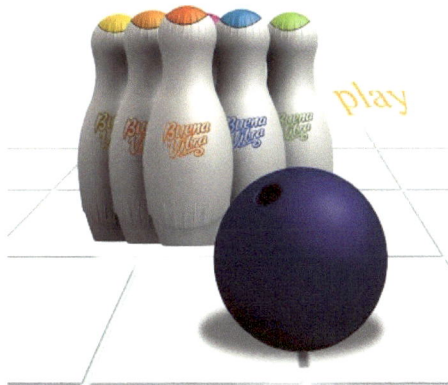

Manufacturer/Designer

Hadas Kruk of Studio Armadillo
Tel fax: 972-03-5298150
Mobile: 972-52-3228211
contact@studioarmadillo.com

- This is a concept packaging designed for a combo pack of 6 detergents with different fragrance / variant.

- This skittle-shaped bottle not only serves as a colorful packaging for different cleaning detergent, but can also be re-used as a bowling game set, including a ball.

- Each detergent has a distinctive cap color.

- The skittle-shaped bottle is very comfortable to use as it allows a perfect grip and pouring by the consumer.

- After use and rinsing, the bottles serve as a real bowling set.

- The packaging is attractive and colorful and adds an additional value of sustainability to the product.

- May be targeted for kids detergent or even normal, but the pack is sure to catch consumer's attention when placed on the shelf.

- Also since the packaging is resuable it reduces waste.

Features:

- Reusability
- No / zero waste packaging

Manufacturer/Designer

London studio
Tel: +44 (0)20 8735 1111
contact@pdd.co.uk
www.pdd.co.uk

- A designer team developed this new concept for PDD's (integrated design & innovation Consultancy) detergent.

- The product is in a tablet form such that three tablets make the equivalent of 1 litre of normal cleaning detergent.

- The packaging is such that the cleaning system includes 1 pump closure, an empty pump and 3 tablets.

- It is compact for storage and lightweight for transporting but lasts as long as a current 1 litre pack.

- Bottle is designed with a pump that pumps detergent after dissolution of concentrated tablet.

- In this concept, there are two possibilities to produce a bottles – one for single handed operation with pump and other one a squarish type which may be used for large volume.

- The design is very innovative and gives fresh look to the product category.

- This may be produced in clear PP (polypropylene) with HDPE pockets or extensions if required.

Features:

- Elegant design
- Easy dispensing by simple pumping

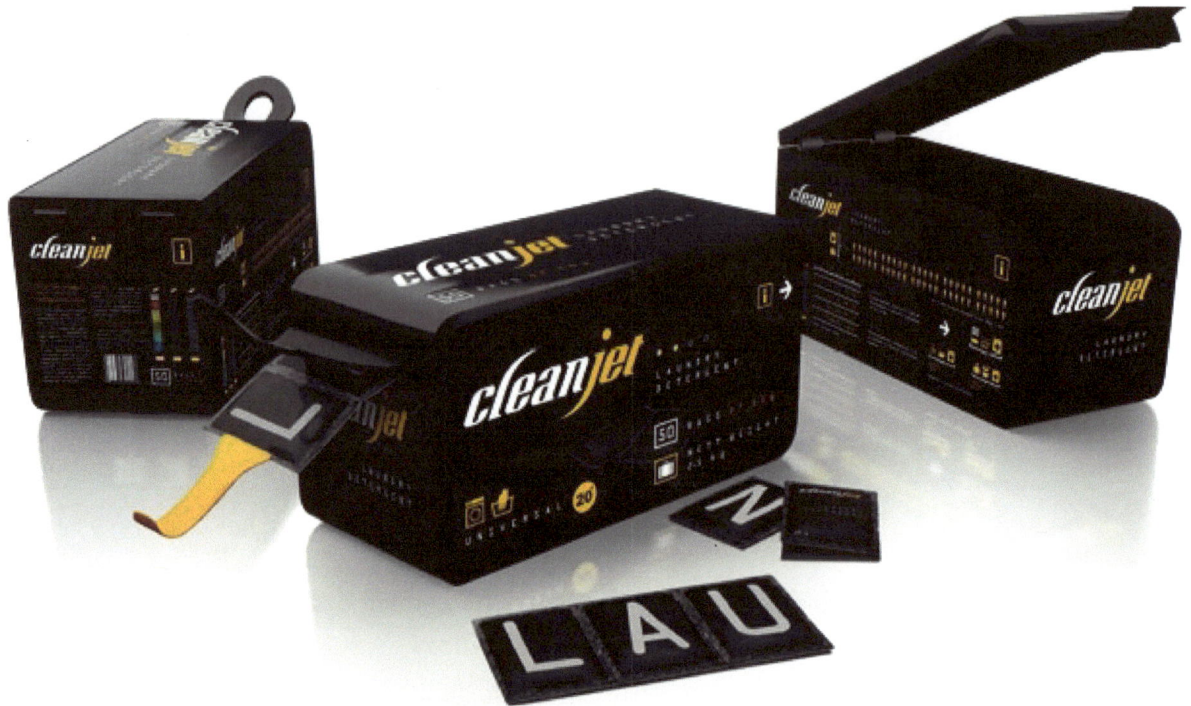

DETERGENT PACKAGING INNOVATIONS

Manufacturer/Designer

Pavla Chuykina Russia
Tel:NA
pavla_chp@mail.ru
website: NA

- This is a concept for packaging sachet string of detergents.

- Pre-measured Sachets in the roll form are packed inside the dispensing box.

- The roll of sachet would move ahead making their way outside the dispenser as and when customer tears off one sachet and the remaining ones will be intact inside the dispenser.

- An additional creative idea is that the graphics of sachets are in such a format that sachets has a special printing one side through which it gives brand letters. Together it may form the name of the brand.

- The dispenser is made of a plastic materials probably PP (polypropylene) or PET (polyester) and this contains around 50 sachets which is equivalent to 50 wash (all-purpose detergent).

- The artwork and design is given a black colour to give it a masculine image and to showcase product power against tough stains.

- This also gives a chance for customer engagement since each sachet has a unique letter printing through which customer may make their own name or any other word.

Features:

- User engagement
- Creative idea to pack and dispense sachets

DETERGENT PACKAGING INNOVATIONS

Manufacturer/Designer

Display Pack Ltd
Tel: 49505/616-451-3061
info@displaypack.com
www.displaypack.com

- The product consists of a solid ball of detergent which is placed into a blue ball dispenser and the product is called as "Toss-n-Go".

- The detergent ball is packed in a thermoformed clamshell inside a conical paperboard sleeve which is equally innovative.

- Most unusual is that the detergent is filled as a hot slurry directly into the clamshell during manufacturing before it cools and hardens and spreads equally in the clamshell blister area.

- The packaging has to be such that seal needs to be weak enough to accommodate the expansion of hot product till it cools down, Yet it has to be strong so that clamshell can be effectively sealed afterwards.

- For this purpose the seal is designed with such a technology that it gets stronger as product cools and expands, but is not be strong so as to be an opening problem for consumers.

- The blue ball having concentrated detergent slurry will be placed directly in the machine for laundry.

Features:

- Creative product design and packaging
- Innovative technology of special kind of sealing

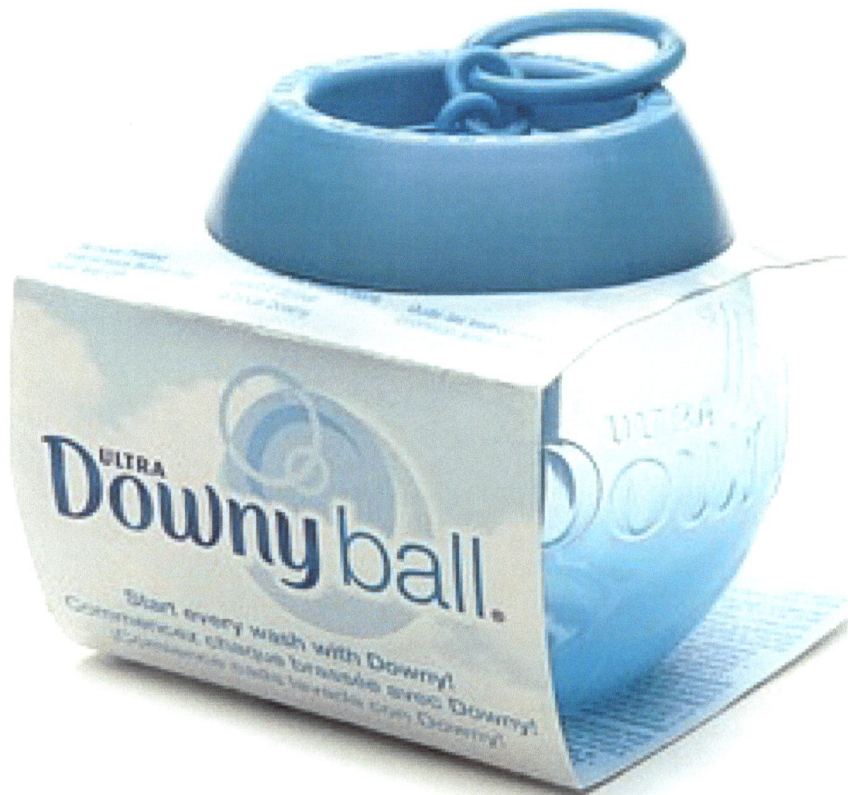

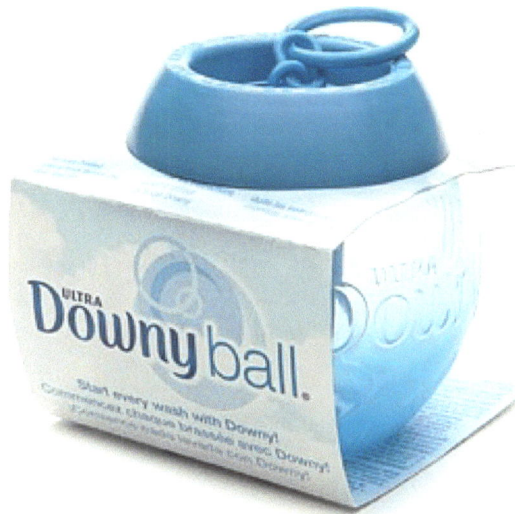

Manufacturer/Designer

P & G USA

Tel: 1-800-688-7638

Email: NA

www.downy.com

- The product is in form of a ancillary dispensing ball (pack) to dispense fabric softener into wash cycle.

- The brand of P&G- "Downy Ball" is a fabric softener which is automatically released during the rinse cycle from the ball.

- The product is such that it doesn't need any outer packaging except a sleeve for branding and labeling.

- Customer have to just pour the Downy fabric softener of their choice up to the right fill line in Downy bottle cap and pour into the ball, seal it, and drop it in at the start of the wash on top of the laundry.

- Easy to pour and control fabric conditioner during wash cycle.

- This idea may be used for promotional packaging of detergent brands wherein this ball replaces measuring cup or scoop.

Features:

- Product itsself serves as packaging
- Hassle free washing

DETERGENT PACKAGING INNOVATIONS

Manufacturer/Designer

Zip-Pack worldwide
Tel:+31 6 512 66 705
Email:paul.iserief@zippak.com
www.zippak.com

- This is an innovative zipper closure from Zip-Pack.

- The Fragrance-Zip is a new zipper closure solution designed to emit a customized aroma upon initial and subsequent openings of a flexible resealable package.

- The scent is embedded into the closure during manufacturing process.

- Zip-Pak can duplicate virtually any desired aroma, enabling a package to replicate a desired scent whenever opened.

- Can be used for detergent pods, chocolates, confectionary, cosmetics, personal care and other sectors.

- This versatile fragrance option can be incorporated into any style of resealable Zip-Pak closure.

- This gives customer a freedom to select the product with its fragrance.

Features:

- Innovative zipper packaging
- May be used for versatile products packed in flexible pouches

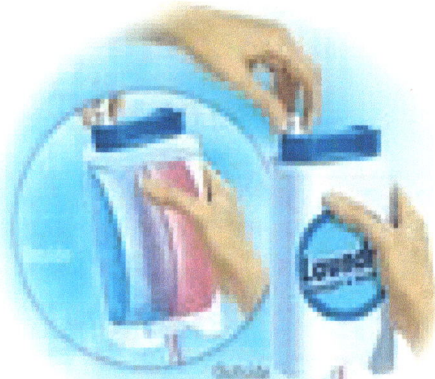

Lastly, "Turnout" the desired amount of product directly over the washer.

A see-through view reveals the simplicity of technology. The wedge activation on the shaft, forcing liquid product to exit.

An alternative solution is to operate the Turnout dispenser by dispensing an electronic button, similar to the concept of an Electric toothbrush.

The Components:

1 **The Reversible Turnout Caddy** accepts both the product pack and the cover. The blue wedge activates downward inside the caddy against the product pack, dispensing the product in measured amounts.

(A) **Typical Consumer Product Pack** Just place pack into caddy. Product refills are virtually instant.

2 **The Caddy Cover** with handle and product view panel. The energy to move the spindle can be **hand-powered** or a **small servo-motor**.

Manufacturer/Designer

Doug Wiesner
Tel: 513-515-0003
doug@turnoutdispensing.com
turnoutdispensing.com

- This is a new green dispensing technology that can significantly reduce landfill waste as well as shipping, storage and packaging costs for manufacturers.

- The 'Turnout' dispenser consists of a patented reversible caddy, which is activated manually or by automatic push button.

- It has no spring or sealed pressure type system.

- The caddy is placed with flexible pack of detergent, upper handle is twisted up to 360degree depending on the quantity desired from the bottom orifice.

- A pack could contain a single product or have dual compartment product packs and is possible to offer both equal 50/50 product ratios, and or unequal product ratios.

- Laundry products could be dispensed automatically into the wash cycle, for instance.

- Dispenser can be either handheld or integrated into an appliance-unit.

- This gives a really innovative idea to dispense any liquid product in a controlled amount.

- Also it provides a freedom of touch free handling of the product.

Features:

- Touch free dispensing of product
- May be used for single or combo product packs like detergent and fabric softener together

DETERGENT PACKAGING INNOVATIONS

DISCLAIMER

- The contents of this book are solely based on the research conducted by PackagingConnections on various innovations available for detergent packaging across the globe.

- PC takes no responsibility for any figures / data presented by the respective manufactures.

- The innovations belong to respective companies.

- This presentation by PackagingConnections (PC) contains information that are provided is accurate to best of our knowledge

- The information provided by the PC in this presentation is general in nature and is not intended as a guide to individual concerns.

- Unauthorized attempts to modify any information stored on this presentation or to use them for any purposes other than its intended purposes are prohibited.

- PackagingConnections makes no warranties, guarantees, or representations as to the accuracy of information contained in this presentation, and assumes no liability or responsibility for any errors in the content.

- Certain names and logos are trademarks and service marks of PackagingConnections and its clients may not be used without permission. Product names, logos, brands, and other trademarks featured or referred to within the presentation are the property of their respective trademark holders.

Dear Readers!
Thank you for your interest.

We are capable of providing customized reports and surveys.
For further details contact:

chhavi.goel@packagingconnections.com

CONTACT US

CORPORATE OFFICE

Sanex Packaging Connections Pvt. Ltd.
(ISO 9001:2008 certified company)
117, Suncity Business Tower
Golf course road, Sector--54
Gurgaon-122002
India
Tel: +91 124 4965770
Fax: +91 124 4143951

Email: info@packagingconnections.com
Website:www.PackagingConnections.com

www.ingramcontent.com/pod-product-compliance
Lightning Source LLC
Chambersburg PA
CBHW041729210326
41598CB00008B/826